About the Author

Germaine Jenkins is a farmer, speaker, wife, and mother. She is the Chief Farm Officer of Fresh Future Farm, an urban farm, grocery store, and event space in North Charleston, SC. She graduated from Johnson & Wales University in 2002 with degrees in Baking & Pastry Arts and Food Service Management. After experiencing the challenges of food apartheid, she decided to do something about it. That's when Fresh Future Farm was born.

You can learn more about Germaine Jenkins and Fresh Future Farm at *www.freshfuturefarm.org*.

Fresh Future Farm Inc.
2008 Success Street
North Charleston, SC 29405
www.freshfuturefarm.org

Correspondence, inquiries:
PO Box 22194
Charleston, SC 29413

This book is intended to supplement, not replace, the advice of a trained health professional. If you know or suspect that you have a health problem, you should consult a health professional. The author and publisher specifically disclaim any liability, loss, or risk, personal or otherwise, that is incurred as a consequence, directly or indirectly, of the use and application of any of the contents of this book.

Copyright © 2019 Germaine Jenkins

All rights reserved.

Published by Fresh Future Farm Inc. 2019

North Charleston, South Carolina

No parts of this publication may be reproduced, stored in a retrieval system, or transmitted in any form or by any means, electronic, mechanical, photocopying, recording, or otherwise, without the prior written permission of the copyright owner.

This book is sold subject to the condition that it shall not, by way of trade or otherwise, be lent, resold, hired out, or otherwise circulated without the publisher's prior consent in any form of binding or cover other than that in which it is published and without a similar condition including this condition being imposed on the subsequent purchaser. Under no circumstances may any part of this book be photocopied for resale.

Cover and interior design by Anik Hall (anikdesign.net)
Edited by Stacia Murphy, Katie Dahlheim, and Kennae Miller

ISBN: 978-1-7344100-0-6

Cookin' Jones

Ignite your Culinary Spirit

Germaine Jenkins

FRESH FUTURE FARM
North Charleston, South Carolina

How to Use this Book

Throughout this book you'll see helpful icons that identify characteristics of each recipe. Most of the recipes teach how to cook from scratch. They are culturally-relevant, low-sodium, and contain lots of veggies. (You may need to invest in some magazines for your bathroom.)

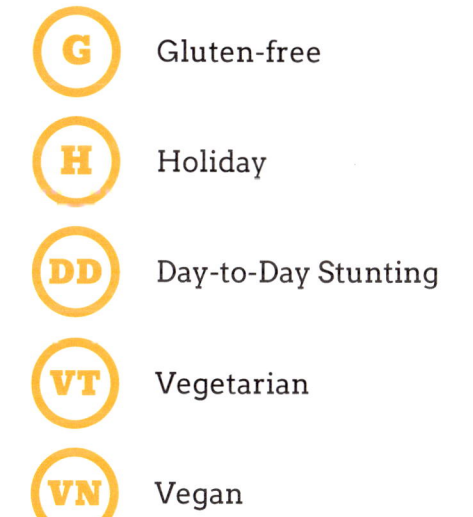

- **G** — Gluten-free
- **H** — Holiday
- **DD** — Day-to-Day Stunting
- **VT** — Vegetarian
- **VN** — Vegan
- **30** — 30-minutes or less

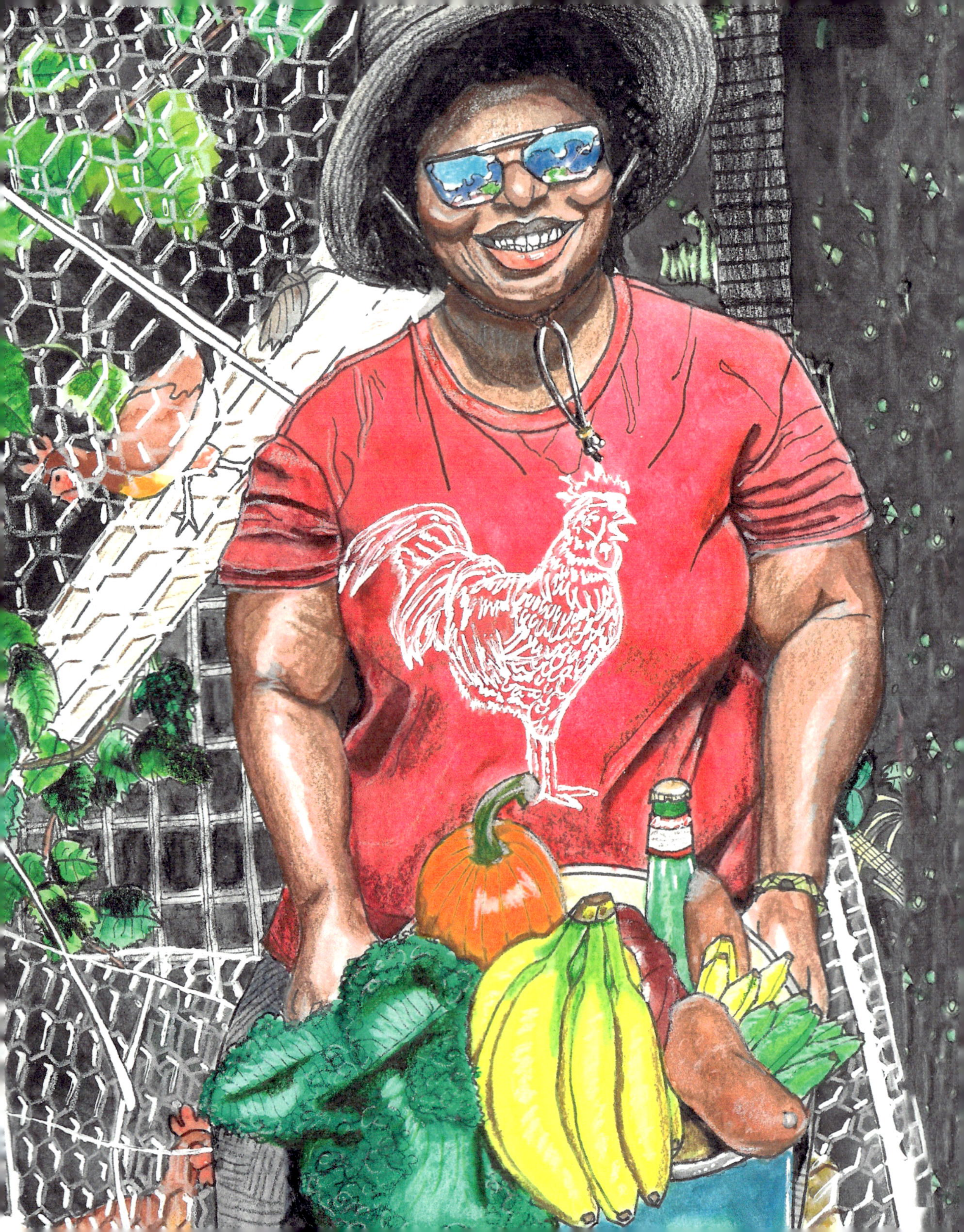

Dedication

So many people come to mind when asked to write a dedication for my first cookbook. Shout out to my late father Willie, who was a brilliant cook on his own. I write this and remember the delicious crockpot roasts he made and him introducing my brothers and me to the concept of a roast beef sandwich. Thanks Daddy! 8D

My mom was a Master Chef to me growing up (except that one time she boiled those rutabagas and those mandarin orange holiday cakes - YEESH!). She was also the person to introduce me to the church cookbooks - where that random mandarin orange recipe lived. You'll see her influence throughout this book.

My children gave me the courage to step beyond my fear and fight for their future. They also were gracious enough to tell me which dishes they didn't care for when they were kids (my mom just got the news last month lol). My dear husband is one of the key reasons that Fresh Future Farm went from being an idea that Todd Chas and I cooked up to a tangible space. Born and raised in downtown Charleston, he gracefully endured 12 years of me trying and failing to use whole grains to make red rice.

Much love to our farm team past and present - Tamazha, LaToya, Anthony, Adrian, Anik, Kenya, Brielle, Joanne, Akua, Tyrell, Jerry, and Brayan. Somehow or another this crew gave me the encouragement, support, motivation, and space to sit still long enough to transcribe the recipes I was already sharing with customers in a form that will be around long enough for the Glam-Baby to put to use.

Last but not least, we are so thankful to our corporate donor HDR for funding this project. Because of their support we can give 100 copies away for free along with the paid and digital versions. Cooking Jones is a term I used to say ahead of cooking up a culinary storm. My hope is that our customers and friends are inspired to be adventurous in the kitchen too.

Contents

Breakfast..1
An Exchange Student's Dream... 2
Raru's Granola..3
Sweet Potato & Pumpkin Pancakes..4
Memories of Brazil..5
Banana Breakfast Shake...6
Stuffed French Toast...7
Sweet Potato Hash..8

Lunch...9
Lettuce Wraps...10
Dilled Tuna & Egg Salad...11
Flax Oat Rosemary Pizza Crust..12
Baked Potato w/ Broccoli..13
Don't Say Nothing Bad 'bout Mustards... 14
Vietnamese Spring Rolls..15

Dips & Sauces..16
Yogurt Dip...17
Balsamic Mustard Vinaigrette..17
Teriyaki Sauce..18
Ranch Dressing..18
Chimichurri - The Miracle Sauce!...19
Chimichurri..20
Chimi Mayonnaise..20
FFF Pesto..21

Sides.. 22
Brazilian Barbeque Dressing.. 23
Brazilian-Style Greens..24
Southern Cabbage Broccoli Slaw...25
Cauli-Mashed Potatoes..26
Healthy Yellow Rice..27
Candied Butternut Squash.. 27

A Holiday Favorite............28
Mom's Cornbread Dressing............29
Sausage Rice Stuffing............31
Kimchi Fried Rice............32

Entrees............33
Mexican Lasagna............34
Squash Tacos............35
Pumpkin Curry............36
Quickie Chicken Cacciatore............37
Chicken Soup............37
African Red Curry............38
Coconut Curried Shrimp............39
A Well-Kept Secret............40
Red Rice for 'Comeyas'............41
10 Veggie Spaghetti............42
Lowcountry Steamed Crabs............43
Cookin' Youths............44
Adrian's Chunky Chili............45
Salmon Stew............46
French Pork n' Beans............47
Quickie Cassoulet............48
A Love for Cooking............49
Arroz con Chorizo............50

Veggies............51
Roasted Broccoli............52
Lowcountry Soup Bunch............53
Okra Soup with Oxtails............54
Teriyaki Eggplant with Spaghetti Squash............55
Chef BJ's Eggplant Creole............56
Veggied Lima Beans............57

Desserts............58
Dump Cake............59
Sweet Potato Bread Pudding............60

Vegan Mousse 61
Coconut Pecan Cookies 62
Peach Charlotte 63
Teas & Tonics 64
Lavender Lemonade 65
Mountain Mint Tea 65
Fresh Herb Sun Tea 66
Ginger Tea 66
Peach Tree Dance 67
Peach Pit Tea 68
Hibiscus Tea Drink 69
Bissop, Sorrel, or Hibiscus Tea 70
Allergy Brew for my Eldest 71
Anik's Kombucha 72

Breakfast

These breakfast recipes reflect the foods I prepare when I have a few minutes:

- **My usual breakfast**: Baby spring greens, spinach, chickpeas, cucumbers, olives, and fresh mushrooms topped with a dollop or two of chicken or tuna salad (no dressing)

- **Healthy on-the-go egg plate**: Scrambled or fried eggs served with any veggie leftovers (kale, sauteed squash or zucchini, lima beans, etc.) as a side dish. Throw some salsa on the eggs and a half an avocado, and you're starting your day great!

Speaking of eggs...the Frugal Gourmet on PBS, taught me that adding about a tablespoon of water keeps the eggs moist and fluffy.

Add a mashed sweet potato to your dropped biscuit mix. Hot biscuits with warmed cinnamon-spiced applesauce - one of my all time childhood favorites my mom made for us as kids in Cleveland. Speaking of biscuits, do yourself a favor and place an upside pan under your next batch when baking. They called that "double panning" in cooking school, and it helps keep those bottoms golden brown and not dusty burnt.

Liver and grits are the truth...

If I want to do something 'brunch-y' on a Sunday morning or to celebrate a birthday, the recipes in this section are some of my favorites.

An Exchange Student's Dream

For ten months in 1988, I lived as a Rotary Exchange student in *Sao Paulo, Brazil*. The only full meal was lunch. Dinner was usually leftovers and breakfast was a cup of coffee and a handful of crackers. The richest smoothest, most delicious coffee I've ever had in my life sat in a thermos for *cafe com leite* after long days going to Universitario, playing volleyball at the club the family joined and hitting the gym (dang I was fit).

I went back for 10 days of missionary work in 2009. Don and Betty's housekeeper was very skilled in the art of American breakfast, and I was all the way confused about pancakes and other standard breakfast fare. Raru's variation of granola was the showstopper for me, and the first thing I recreated when I got back to the States. This cookbook is filled with recipes that all of the distinctive taste buds in my house enjoy. This toasted oat concoction is one of the favorites. When we slept on a boat near an Amazon River village, daydreams of Raru's granola-topped cooked oatmeal made sleepless snore-filled nights bearable.

A fresh batch of granola coming out of the wood oven at the farm.

Raru's Granola
Serves 12

¼ stick of butter
3 cups old fashioned oats
1 cup chopped nuts (walnuts, almonds, pecans, or Raru's secret - Brazil nuts)
¼ cup honey

cinnamon to taste
butter cooking spray
½ shredded coconut
½ cup raisins or dried blueberries
9 ½" X 13" baking pan

Preheat oven to 350°F. Melt butter in a baking pan in the oven. Remove pan from the oven and toss oats and nuts in melted butter. Drizzle mixture with honey and butter cooking spray. Stir and return to oven. Stir every 10 - 15 minutes to keep oats and nuts from burning. After about 40 minutes, stir in coconut and raisins. Then, turn off the oven. Allow granola to set in oven for another 30 minutes. Remove and let cool. Store granola in an airtight container or gallon storage bag in the refrigerator.

Try Raru's granola over hot cooked oatmeal - OTHERWORLDLY DELICIOUS!

Sweet Potato & Pumpkin Pancakes

Serves 8 - 10

3 mashed sweet potatoes OR canned pumpkin
3 Tbsp molasses
½ cup flax seed
½ cup old fashioned oats

cinnamon, to taste
1 dash nutmeg (optional)*
3 cups of pancake mix
1 or more cups of water (for your desired thickness)

Notice the order that the ingredients are mixed. We usually deal with the wet ingredients first - minus the water - before mixing in the dry stuff. Over-mixed pancake can be tough like flip flops. Add enough water to the mix to get the consistency a little thicker than buttermilk, The quantities are large because we like to make enough for extra pancakes during the week. They freeze well too!

NOTE: *The nutmeg really enhances the canned pumpkin.*

Before he was a farmer, little Adrian was my sous chef in our homeschooling days

Memories of Brazil

Another recipe that has its roots in Brazil. I found and fell in love with juice bars in the *Sao Jose do Rio Preto* town that was my home. If I wasn't scarfing down passion fruit juices, I was on that banana juice tip. Banana juice was really ripe bananas blended with milk.

I modified the recipe. When I noticed my husband losing weight early in our marriage to be more of a protein shake. The simple addition of nonfat dry milk (powdered goat milk) mixed with vanilla pudding mix makes a delicious homemade version of an instant breakfast. Freeze peeled bananas (peaches or berries), and it comes together in a few minutes.

Thirty nine year old me in the port city of *Santarem*, Brazil translating for my mission trip team and wondering what my fresh future back in North Charleston would look like

(G) (VT) (VN)

Banana Breakfast Shake
Serves 2

1 frozen banana
1 Tbsp homemade breakfast mix
1 cup milk
blender

¼ cup coconut cream (optional)*
¼ cup coconut milk (optional)*
dash of vanilla extract (optional)*

Combine all ingredients into a blender and blend until smooth. Banana can be substituted with different fruits, such as strawberries and peaches.

NOTE: *Make vegan by substituting milk for ¼ cup coconut cream and ¼ cup coconut milk with a dash of vanilla.*

A farm demo and video with dairy and non-dairy versions of this breakfast treat!

Stuffed French Toast
Serves 4

8 slices Texas toast or sourdough bread cut thick
½ cup seedless strawberry, blackberry or raspberry jam
5 eggs
½ cup almond or coconut milk
1½ teaspoons vanilla extract
1 tsp cinnamon
4 tablespoon butter
cooking spray
baking sheet
serrated or steak knife

Preheat oven to 350°F.

Use a knife to cut slits in the bread. Use a spoon to carefully add a teaspoon of jam in each slit. Place stuffed bread slices in a buttered baking dish.

In a medium bowl, beat together the eggs, almond milk, vanilla extract and cinnamon. Pour the egg mixture over the stuffed bread. For best results, refrigerate overnight.

In a large heavy skillet over medium high heat, melt butter. Place two pieces of toast in the skillet, and cook on all sides until golden brown.

COOKING SUGGESTION:
This recipe would be FIRE cooked on a greased waffle iron. Bougie breakfast level goes to 12.

TIP: *You can use a fancy serrated knife or a steak knife to slice a loaf of sourdough bread.*

Substitute eggnog and apricot preserves for a holiday batch of french toast.

Sweet Potato Hash
Serves 2

1 sweet potato diced
butter or olive oil
chili powder, to taste

salt, to taste
a dash of Italian seasoning blend
saute pan

Heat the pan to medium heat. Add a small amount of butter or a drizzle of oil. When hot, add the sweet potatoes. Toss or stir every 5 minutes until brown. Add chili powder, salt, and Italian seasoning.
Serve and enjoy!

You can easily add onions and bell peppers to this recipe. We kept it really simple here.

Lunch

After spending time at the Women's Environmental Institute in North Branch, MN, I had high hopes of bringing a communal eating situation to the farm. Unfortunately, some goals become complicated when you don't live on site. Anyway, here are some quick lunch options that you can prepare ahead and feel like you did something special for yourself and your family.

Women's Environmental Institute In North Branch, MN

Lettuce Wraps
Serves 8

Too many starches in your life or thinking of a way to sneak more veggies in? Replace tortilla shells with bib or butter crunch lettuce. These pack nicely the night ahead for lunch.

Version 1:
- bib or butter crunch lettuce
- rotisserie chicken strips
- bell pepper slices
- fresh cilantro
- garlic greens
- chimi mayo or chimichurri

Version 2:
- bib or butter crunch lettuce
- 1 Tbsp tuna salad
- green onions
- garlic greens
- match stick-cut cucumbers
- Chimi mayo or chimichurri

Place the desired fillings on top of a slice of lettuce with one of the chimi sauces. Wrap and enjoy or pack for the next day's lunch! No soggy bread in your lunch bag!

Chicken salad, rotisseries chicken and tuna salad wraps with farm fresh herbs!

Dilled Tuna & Egg Salad

Serves 6

Easy way to pump up your tuna or egg salad is adding a teaspoon or two of fresh chopped dill. Now you fancy! Sauteed and cooked kale and mushrooms for all the vegans!

2 (5 ounce) cans tuna, drained and rinsed
3 hard-boiled eggs, peeled and chopped
½ stalk celery, chopped
⅓ cup mayonnaise
2½ teaspoons sweet pickle relish or cubes
1 tbsp fresh dill, chopped

Combine tuna, eggs and celery in a medium sized bowl. Stir in mayonnaise and dill. Serve at room temperature or chill until ready to serve.

Chicken or Turkey Salad - substitute 10 oz. chicken or turkey for tuna with celery, mayonnaise, relish, and 1 Tbsp honey or stone ground mustard. Instead of cubing cold chicken or turkey, crumble meat between fingers before mixing with remaining ingredients. One of my mom's favorite post-Thanksgiving treats was my turkey salad.

Egg Salad - leave out the meat and add 1 tbsp honey or stone ground mustard

Farm fresh eggs sing when you add fresh dill.

(DD) (VT) (VN)

Flax Oat Rosemary Pizza Crust
Makes 4 medium pizza crusts

6 cups all purpose flour
¼ cup fresh rosemary
½ cup old fashioned oats
2 Tbsp ground flax seed

1 packet of active dry yeast
3 Tbsp olive oil, optional
3 cups water, room temperature

Mix first 6 ingredients in a large bowl. Stir to combine. Add 2 cups of water and stir ingredients with a fork until the dough starts to pull together. Add the last cup of water and finish stirring until it turns into a ball.

Cut the dough ball in half and place each half in olive oil coated resealable bags so the dough could rise without overflowing and easily be transported to parties or the freezer!

PRO TIP: *Tear string cheese into thinner strips for stuffed crusts!*

In a pinch, you can mix oats, flax, and rosemary into frozen bread dough that thaws overnight.

Baked Potato w/ Broccoli
Serves 1

1 medium baking potato
1 teaspoon olive or avocado oil
Pinch of salt

1 cup roasted broccoli (hot or cold)
¼ cup ranch dressing
Cayenne pepper to taste

Preheat the oven to 300 degrees. Scrub the potato, and pierce the skin several times with a knife or fork. Rub the skin with oil and sprinkle with salt. Bake potato for an hour or until soft and golden brown. Slice the potato down the center, add the roasted broccoli, ranch dressing and sprinkle with cayenne pepper to taste. Substitute ranch with a chimichurri drizzle for a vegan version.

I used to poke holes in the potato and microwave for a healthy lunch in minutes. Urban Growers Collective culinary herb salt is a spice-free substitute for cayenne.

Don't Say Nothing Bad 'bout Mustards

One of my favorite winter greens are mustards. I was standing in my backyard, snacking on raw greens one fall when this recipe came to me. Fresh mustards have a peppery taste similar to wasabi. As I stood there chewing, I wondered what would happen if I added it to a rice paper spring roll. Along with lots of fresh veggies, pickled ginger and shrimp give this fancy appetizer some extra oomph.

I made these first for a wedding anniversary for hubby.

Vietnamese Spring Rolls
Serves 5

10 rice wrappers
2 ounces rice vermicelli noodles
3 Tbsp fresh parsley
3 Tbsp fresh cilantro
10 small leaves fresh mustard greens
10 small leaves lettuce
1 small cucumber, halved, seeded and julienned
10 shrimp, peeled deveined and cut in half
20 pieces of pickled ginger

Bring a medium saucepan of water to boil. Boil rice vermicelli 3 to 5 minutes, or until al dente, and drain.

Fill a large bowl with warm water. Dip one wrapper into the hot water for 1 second to soften. Lay wrapper flat. In a row across the center, place 2 shrimp halves, a handful of vermicelli, basil, mint, cilantro and lettuce, leaving about 2 inches uncovered on each side. Fold uncovered sides inward, then tightly roll the wrapper, beginning at the end with the lettuce. Repeat with remaining ingredients.

NOTE: *This recipe is COOKING JONES LEVEL 7 because prep time and dirty dishes are the devil. The spring rolls are heavenly!*

Due to prep time, I save these beauties for special occasions

Dips & Sauces

The year I spent homeschooling Adrian, preparing homemade sauces came to me like second nature. Do-it-yourself dips allow you to adapt ingredients to fit your dietary needs - sugar free, soy free, dairy free, no salt - no problem.

Mixing sauces in jars is also a great way to get the youngsters in your home involved in the kitchen without handling sharp utensils.

Homemade and semi-homemade dressings from a North Charleston school nutrition camp.

16

(G) (DD) (VT)

Yogurt Dip
Serves 6

2 cups Greek yogurt
2 teaspoon honey
1 teaspoon vanilla extract
dash of cinnamon

Stir the yogurt, honey, and vanilla extract in a bowl. Chill for two hours. Great with fruit trays! Cashew yogurt is good substitute for vegans.

(G) (DD) (VT) (VN)

Balsalmic Mustard Vinaigrette
Serves 8

½ cup olive oil
¼ cup balsamic vinegar
⅛ cup honey or sugar
2 tbsp regular Dijon or stone ground mustard

Combine the olive oil, balsamic vinegar, honey, and mustard in a glass jar with a lid. Replace lid and shake until thoroughly combined. Store in refrigerator.

NOTE: *This dressing is delicious on pasta salad.*

Teriyaki Sauce
Serves 6

¼ cup soy sauce (substitute coconut aminos for those allergic to soy)
½ cup chicken broth
½ cup rice wine vinegar
1 teaspoon sesame oil
2 tablespoons honey
2 tablespoons fresh (or 2 tsp ground ginger)
4 cloves garlic, minced

In a small bowl, combine ingredients till blended. Then, you'll have a yummy, homemade sauce that allows you to control your family's sodium and preservative intake.

Ranch Dressing
Serves 12

1 cup mayonnaise
½ cup buttermilk
1 tbsp fresh chives, chopped
1 tbsp fresh parsley, chopped
1 tbsp fresh dill weed, chopped
1 clove garlic, chopped
¼ onion, diced small
⅛ teaspoon salt
⅛ teaspoon ground black pepper

In a large bowl, whisk together the mayonnaise, sour cream, chives, parsley, dill, garlic powder, onion powder, salt and pepper. Cover and refrigerate for 30 minutes before serving.

WARNING: *You may wanna slap the bottled version after trying the homemade alternative. Do not blame the author*

Chimichurri - The Miracle Sauce!

I first discovered this sauce while working a catering event at the Charleston Aquarium. It's traditional to serve it as a topping for beef. I started making it for my family and would use it as a healthy salad dressing substitute. One of my uses was tossing a chopped lettuce and pasta salad in chimichurri. Kristin Kirkpatrick, MS, RD, LD, dietitian and wellness manager for the Cleveland Clinic Wellness Institute says that vitamin A improves eye health and Vitamin K is great for blood health. And it's flavorful enough to replace added salt in recipes. Cilantro is an antioxidant that can have an antibacterial effect against salmonella. Tt also adds iron, magnesium, and manganese to your diet. Cilantro is a natural diuretic and can treat nausea.

One time, we grilled steak and shrimp kebabs in our earthen oven at the farm and made a couscous dish in a *tagine*. The only seasonings we used were fresh lime juice and chimichurri as a topping. Everyone who ate was pleased with the flavors to say the least.

Green salad, steak, shrimp kebabs and couscous all dressed with chimichurri!

G DD VT VN

Chimichurri
Serves 10

1 cup olive oil
8 cloves garlic, chopped
⅓ cup red wine vinegar, or more to taste
1 teaspoon salt, or to taste
½ teaspoon ground cumin
1 cup fresh cilantro leaves
2 bunches flat-leaf Italian parsley, stems trimmed

Combine oil, garlic, vinegar, salt, cumin, cilantro, oregano, and parsley in a blender.

Pulse blender 2 to 3 times; scrape down the sides using a rubber spatula. Repeat pulsing and scraping process until a thick sauce forms, about 12 times.

G DD VT

Chimi Mayonnaise
Serves 4

1 cup light mayonnaise
¾ cup fresh cilantro, chopped
¼ cup fresh parsley, chopped
1½ Tbsp lime juice
1 tsp low sodium soy or coconut aminos
⅛ tsp garlic powder

Blend, chill, store in mason jar in the refrigerator. Great on sandwiches, wraps, or as a salad dressing.

G DD VT

FFF Pesto
Serves 12

When I first started working with children over the summer to introduce them to healthier eating, I used pesto with cherry tomatoes and bacon as an introduction to a garden pizza. Years later, I went to an Italian restaurant that served the basil gravy on mashed potatoes - game changer.

3 cups packed fresh basil leaves
4 cloves garlic
¾ cup grated Parmesan cheese
½ cup olive oil
½ cup chopped spinach
¼ cup pecans, toasted

In a food processor, blend basil, garlic, Parmesan cheese, olive oil, spinach and pecans until smooth. Store in the refrigerator.

Young campers made a deliciously rustic pesto for pizza!

Sides

For me, side dishes are golden because they can serve double duty as a side for breakfast, lunch, and dinner. As an example, my Brazilian Style Greens are dreamy served with beans and rice for dinner or eggs and rice for breakfast.

The goal with the sides featured in this cookbook is that you can cook them in 40 minutes or less and impress a crowd.

That feeling when you realize the plain old sauteed veggies from your frigde are a great side on their own.

Brazilian Barbecue Greens

When I was 17 years old, I spent ten wonderful months as a Rotary Youth Exchange Student in Sao Paulo, Brazil. My third host family introduced me to the Brazilian barbecue restaurants called 'churrascarias.' Along with the yummy grilled chicken hearts, pork loins and feijoada, I LOVED those Brazilian styled greens that were sliced ribbon thin, coated in oil and cooked until tender. Whether they're collards, kale, mustards, turnips or a combination of greens, this preparation method gives off a natural sweetness. This four ingredient recipe becomes vegan when you replace chicken broth with water or vegetable broth.

PLEASE NOTE: *This variation of greens does not produce much potlikker. Sorry not sorry. You'll love them just the same.*

This mixed greens variation includes mustards, turnip roots, and turnip greens.

Brazilian-Style Greens
Serves 6

2 pounds collard greens - rinsed, rolled cigar style and sliced into 1-inch ribbons

⅓ cup oil or bacon fat, separated
6 cloves of garlic, minced
4 cups chicken broth

Place a few handfuls of collards in a large pot over medium-high heat. Coat with some of the oil and stir till coated. Repeat process until all the greens are coated. Stir in chopped garlic and two cups of broth. As the broth reduces add more liquid, one cup at a time, stirring occasionally. Cook until greens are tender, about 45 minutes total. Feel free to add more broth and vinegar as desired.

Greens are calcium rich and may support eye health. Just steam the kale before juicing, y'all!

Southern Cabbage Broccoli Slaw
Serves 10

1 head cabbage, finely chopped
2 head broccoli, finely chopped
2 carrots, finely chopped
1 cup mayonnaise

2 Tbsp dijon or stone ground mustard
⅓ cup white sugar or honey
⅓ cup apple cider vinegar

I usually hand chop veggies, but a cheese grater in a large bowl will make this an easier task. Mix cabbage, carrots, and broccoli in a large salad bowl. Whisk mayonnaise, sugar, and vinegar until smooth and the sugar has dissolved. Pour dressing over cabbage mixture and mix thoroughly. Cover bowl, and refrigerate slaw at least 2 hours. Mix again before serving.

Exercise your choppers on this rough chopped broccoli cabbage slaw!

Cauli-Mashed Potatoes
Serves 6

8 red potatoes, skin on, washed and cubed
1 head of cauliflower, green leaves removed, washed and cubed
3 garlic cloves, peeled
1 pinch salt

⅓ to ½ cup butter
¼ cup chicken broth or water
¼ teaspoon poultry seasoning
2 teaspoons of garlic powder
1 teaspoon black or white pepper
2 teaspoons salt

Place the potatoes, cauliflower and garlic cloves in a large pot, and fill with enough water to cover. Bring to a boil, and cook for about 20 minutes, or until easily pierced with a fork or potato masher. Drain the potatoes and keep the boiled garlic. Add the butter, broth, poultry seasoning, garlic powder and salt. Mash until the butter is melted.

NOTE: *These non-traditional toppings take this dish to the next level!*
- *Pesto is an Italian option that I tasted at a restaurant in Portland years ago (made me start calling it basil gravy in my mind).*
- *Caramelized Onions are a lighter easier topping for mashed potatoes, so good!!!*

Fresh or frozen cauliflower kick up the fiber content in your mashed potatoes without losing flavor.

Healthy Yellow Rice

Serves 8

2 Tbsp olive oil
2 cup rice
¼ cup dried minced onion
1 teaspoon ground turmeric

1 teaspoon garlic powder
½ teaspoon ground black pepper
1 teaspoon salt
4 cups hot water

Add rice to a large saucepan over medium heat. Add oil, onion, turmeric, garlic powder, black pepper, and salt. Stir until combined. Remove from heat, and transfer to a baking dish. Add hot water, cover (with a lid or foil), and bake at 350 degrees for 45 minutes to an hour. Fluff rice with a fork.

NOTE: *Add cooked ground turkey for dirty rice.*

Candied Butternut Squash

Serves 4

1 butternut squash, washed and split in half
¾ cup sugar or honey

3 cups water
cinnamon, to taste

Preheat oven to 275°F. Add two inches of water to a baking pan. Place squash halves in pan. Bake at 275°F until skin softens and liquid creates a sauce about 1 hour. Try this recipe in a crock pot on low heat!

NOTE: *A customer taught us that winter squash skin is edible and tasty, and that it's really tasty with sugar.*

A Holiday Favorite

Growing up, the holidays were the times when my mom would go all out. As much as I adore greens and turkey, her two signature dressings, made with cornbread and rice, were the star of the show. While doing interpretive work with Joseph McGill and the Slave Dwelling Project, I learned that what we grew up with as cornbread dressing has roots that go all the way back to West Africa. Kush, short for couscous, was a Islamic influenced dish eaten in West Africa. Enslaved Africans substituted cornbread to recreate a dish that resembled kush. Like my mom, I like a wetter dressing (that way leftovers don't come out dry as a bone), but you can add liquid to achieve the level of moisture you prefer.

Head Cook, Jerome, and I at Magnolia Plantation about to throw down over an open hearth.

Mom's Cornbread Dressing
Serves 8

Step One (Day 1) - CORNBREAD (I usually prepare the cornbread the night before but in a pinch it can be prepared a few hours ahead and used once cooled):

1¼ cups milk or buttermilk
2 cups self-rising cornmeal
2 eggs, beaten
1 teaspoon garlic powder
1 teaspoon ground sage
¼ cup unsalted butter or bacon fat, melted
1 tablespoon oil or bacon fat

Preheat oven to 425°F and set skillet inside to heat up. Place 9-inch cast-iron skillet in oven to warm it. Mix dry ingredients - cornmeal, and spices in a large bowl. Mix in eggs and buttermilk until batter is smooth. Remove skillet from oven. Swirl the butter the skillet to coat; pour off excess. Pour batter into the skillet. Bake in a preheated oven until a toothpick inserted into the center comes out clean, 18 to 23 minutes. Let cool and cover.

Ready for the oven!

STEP TWO (Day 2) - PUTTING THE DRESSING TOGETHER:

¾ cup butter
2 onions, chopped
1 green bell pepper, chopped
2 stalks celery, chopped
1 pound pork sausage
2 teaspoons dried sage
2 teaspoons garlic powder
1 teaspoon dried thyme
1 teaspoon poultry seasoning
1 teaspoon salt
½ teaspoon pepper
½ cup chopped fresh parsley
1 can cream of mushroom soup
1 tablespoon sweetened condensed milk
3 cups chicken stock

Saute chopped vegetables in oil in a large skillet over medium heat. Cook until softened then add ground sausage. Transfer mixture to a large bowl. Crumble cornbread in bowl. Add cream of mushroom soup, seasonings and broth and mix by hand. Add fresh parsley and mix until combined. Transfer to a greased baking pan and bake at 325°F for an hour.

NOTE: *Here are some lowcountry variations to Cornbread Dressing:*
- *Add ¾ cup chopped shrimp or crab meat (do not add sweetened condensed milk if using crab meat as it sweetens the dressing)*

Sausage Rice Stuffing
Serves 6

For Martians who don't like cornbread dressing; you could also think of this as a Holiday Dirty Rice. Delicious with greens!

- ¼ cup bacon fat or butter, melted
- 1 onion, chopped
- 2 stalks celery, chopped
- 1 bell pepper, seeded and chopped
- 1 package breakfast or sage sausage
- 1 teaspoon garlic powder
- 1 teaspoon ground sage
- ½ teaspoon celery seeds
- ½ teaspoon poultry seasoning
- 2 ½ cups rice (white, brown or Jasmine)
- 5 cups broth or water

Saute vegetables in oil over medium heat. Cook until veggies soften, and add sausage. Once sausage is browned, add spices and rice. Stir to coat rice. Transfer to baking dish with lid. Add liquid, and bake at 350°F for 45 minutes to an hour.

Chopped and sauteed mushrooms would kick this rice dish to the next level.

Kimchi Fried Rice
Serves 6

3 Tbsp olive oil, divided
4 cups kimchi, chopped
1 Tbsp butter
4 cups cool cooked white rice, preferably 1 day old
1 Tbsp sesame oil
4 eggs
seasoned seaweed, crushed
4 scallions, chopped
white pepper, to taste
garlic powder, to taste
salt, to taste
sesame seeds

Pour 1 Tbsp of oil into large skillet over medium-high heat. When simmering crack eggs over the skillet. Lightly scramble, leaving the eggs slightly wet. Place in a bowl and set aside.

Clean the egg residue from the pot and put it back on the stove. Pour 2 Tbsp oil into large skillet and turn to medium heat. When the oil is hot, add kimchi. Cook while stirring until kimchi is heated through. Add the butter and stir until incorporated.

Add the rice, breaking up clumps. Stir until it is evenly mixed with the kimchi. Add sesame oil, garlic, and salt. Then stir. Spread rice mixture in an even layer until it gets a little crispy. Add the eggs back in. Turn off heat and divide between bowls. Garnish with crushed seaweed, scallions, pepper, and sesame seeds.

NOTE: *You can make this recipe very quickly. It was Anik's go-to quick dinner. Anik likes to mix in some sesame seeds as well as using them as a garnish. They give a little extra texture.*

Entrees

Tips for moist and flavorful and light boneless skinless chicken breasts:
- Buy individually quick frozen (IQF) breasts
- Rinse desired amount in water (sanitize your sinks after y'all)
- Coat a baking pan with cooking spray
- Add seasonings to pan (garlic powder, onion powder, poultry seasoning, Mrs. Dash, etc.)
- Place chicken on pan
- Spray tops with cooking spray and season tops of breasts
- Bake FROZEN at 350°F for 45 minutes to an hour

1/4 cup each prepared mustard (any kind), brown sugar and Blenheim ginger ale reduced to a glaze

Mexican Lasagna

Serves 6

I'm pretty sure a cooking show taught me that cumin, chili powder, and oregano make a homemade, no-salt version of taco seasoning. This lasagna recipe is great because it reheats and freezes well.

1 pound lean ground beef
2 (16 ounce) cans refried beans
1 medium onion, diced
1 bell pepper
1 tablespoon ground cumin
2 teaspoons chili powder
2 teaspoons oregano or Italian seasoning
1 jar tomato sauce
1 large jar (or 2 regular sized jars) salsa
6 (10 inch) flour tortillas
3 cups shredded Cheddar or Mexican cheese
2 green onions, chopped
1 cup chopped cilantro, optional

Preheat oven to 375°F. In large skillet over medium heat, cook beef, then brown the drain. Add onions and refried beans and cook until warmed through. Stir in seasonings to combine and add salsa. Cook until warmed through. Cover base of baking pan with a thin layer of meat mixture. Cover with 3 overlapped tortillas. Sprinkle with 1½ cups of cheese. Spread half of remaining meat mixture over cheese. Top with remaining tortillas, top with remaining meat mixture and top with cheese. Bake for 45 minutes. Top with green onions and chopped cilantro. Let cool 5 minutes before serving.

Squash Tacos
Serves 3 - 4

For the Filling:
- 1 pound of ground meat (or 1 can refried beans with ¼ cup water or broth)
- 1 can each black beans, rinsed and drained
- Squash seeds
- Half of a chopped onion
- Half of a chopped bell pepper
- 1 teaspoon each ground cumin and chili powder
- 1 teaspoon garlic powder
- ½ teaspoon dried oregano
- Half cup salsa
- Shredded cheddar or Mexican cheese (optional)
- Green onions, optional
- Cilantro, optional, for garnish

For the Tacos:
- 3 yellow squash or zucchini halved lengthwise, seeds scooped out and set aside
- 2 tablespoons oil or cooking spray

Preheat oven to 400°F, coat squash halves in oil and roast for 30 minutes, until they are tender, set aside and cool.

Brown ground meat (or drained black beans, refried beans and ¼ cup water) and add onions, squash seeds and meat, spices and simmer for about 10 minutes. Fill squash halves with seasoned meat or bean mixture. Top with salsa and cheese and bake for 10 minutes. Remove from oven and garnish with green onions and cilantro.

NOTE: *Fill with refried beans and sauteed kale for vegan tacos.*

Pumpkin Curry
Serves 6

1 can of pumpkin or ½ of a baked pie pumpkin
1 cup onions, chopped finely
1 green chili
1 tsp ginger, grated or ginger paste
1 to 1 ½ tsp coriander powder
½ to 1 tsp garam marsala or any curry powder (we prefer both)
½ tsp red chili powder
⅛ tsp turmeric
Salt as needed
2 Tbsp oil
½ tsp cumin

Heat oil in a pan. Add cumin and mustard until they start to sizzle. And ginger and fry together until the ginger is fragrant. Add onions and green chilies. Fry until they turn golden. Sprinkle salt and turmeric. Then saute for 2 minutes. Add chili powder, garam marsala, coriander powder, and red chili powder. Fry for 1 to 2 minutes. Add pumpkin and cook on low until it is fully immersed. Keep covered for 10 minutes so that the flavors are absorbed. Serve with rice and enjoy!

NOTE: *Anik came up with this recipe in an attempt to calm their reflux. The only tomato-like subsitute we had was canned pumpkin for a pie.*

Delicious curry without the reflux

Quickie Chicken Cacciatore
Serves 6

3 tablespoons oil
1 rotisserie chicken
5 cloves of garlic
¼ - ½ cup balsamic vinegar

Divide chicken into pieces with skin. Saute in olive oil until warmed through. Stir in garlic and balsamic vinegar.

Chicken Soup
Serves 8

4 tablespoons oil
1 large onion
2 stalks celery
1 large carrot
1 bunch kale, chopped
4 garlic cloves, chopped
1 small piece of ginger, fine chopped
2 tablespoons of garlic powder
2 teaspoons ground turmeric
1 teaspoon poultry seasoning
Salt and pepper, to taste
Leftover rotisserie chicken, especially the dark meat portions with the bone
1 quart water or broth
1 can fire roasted tomatoes or 2 large tomatoes, chopped
1 cup rice or pasta

Saute onions, celery and carrots, and chopped kale in oil. Add garlic cloves and ginger. Stir in garlic powder, turmeric, thyme, poultry seasoning and black pepper. Add drumsticks, and thighs (skin removed with bones) and a can of fire roasted tomatoes to pot. Add 1 quart of broth or water and bring to a boil. Add breast meat, 1 cup of rice or pasta and simmer for another 20 minutes. Add salt and pepper to taste.

African Red Curry
Serves 6

1 rotisserie chicken
1 sliced pepper
2 large onions
2 large red peppers
1 stalk of lemongrass
2 Tbsp peanut butter

1 tsp fish sauce
1 chopped sweet potato
1 handful of basil
1 cup of coconut milk
salt, garlic powder, curry powder, and coriander, to taste

Preheat oven to 400°F. Chop veggies, toss them in oil, and place them in baking dish. Season with salt, garlic powder, curry powder, and coriander. Add coconut milk, peanut butter, and stir. Break chicken down and submerge in your veggie and liquid mixture. Bake for an hour and a half. Add basil for garnish. Serve over rice of your choice.

FUN FACT: *Michael Twitty told me basil makes chicken sing - FACTS!*

Adding the larger bones in as you simmer create so much flavor.

Coconut Curried Shrimp
Serves 4

1 can of coconut milk
curry powder, to taste
A dash or Thyme
garlic powder, to taste
Loisa Organic Adobo, to taste

white/black pepper, to taste
cilantro, chopped
shrimp, peeled and deveined
sea salt, to taste

Heat pan to medium. When hot, add coconut milk and a generous amount of curry powder, thyme, garlic powder, white pepper, and sea salt. Let the sauce boil until it turns brownish-yellow. Then, add a handful of chopped cilantro. Stir in the cilantro and let sit for about 5 minutes. Add shrimp (cooked shrimp saves time, but raw shrimp better absorbs the flavor of the curry). When the shrimp is heated or pink, you're done! Serve with rice and pink beans. Enjoy!

NOTE: *This recipe came to us after my eldest got married. #whenyoursoninlawcomesthrough #caribbeanvibes. He prefers the Thai coconut milk to the Goya brand.*

Last time they made the dish, they cooked the shrimp in the air fryer first - YUM!

A Well-Kept Secret

For a *comeya* like me who wasn't aware of the many downtown Gullah eateries, the best plate of red rice in downtown Charleston was at the old Piggly Wiggly on Meeting Street. Those senior women could cook! Lines wrapped around that deli counter, and you had better have your order figured when it was your turn. I failed at recreating their rice masterpiece for 16 consecutive years. Brown rice doesn't work, stirring makes it gummy but my dear husband sat through every failed batch. If you love red rice and can't afford to go to Hannibal's restaurant all the time, this dish is for you. Just know that there are 10,000 other ways to make red rice, and you aren't ready. I finally watched a couple YouTube videos to help me figure out the error of my ways, I came up with this 'red rice for dummies.' The variable here is the rice. Use the parboiled Uncle Ben's and hush. Pour till it looks like this picture and hush. Cover the dish with foil and HUSH! Don't be experimenting until you've got good results with this one at least 5 times in a row.

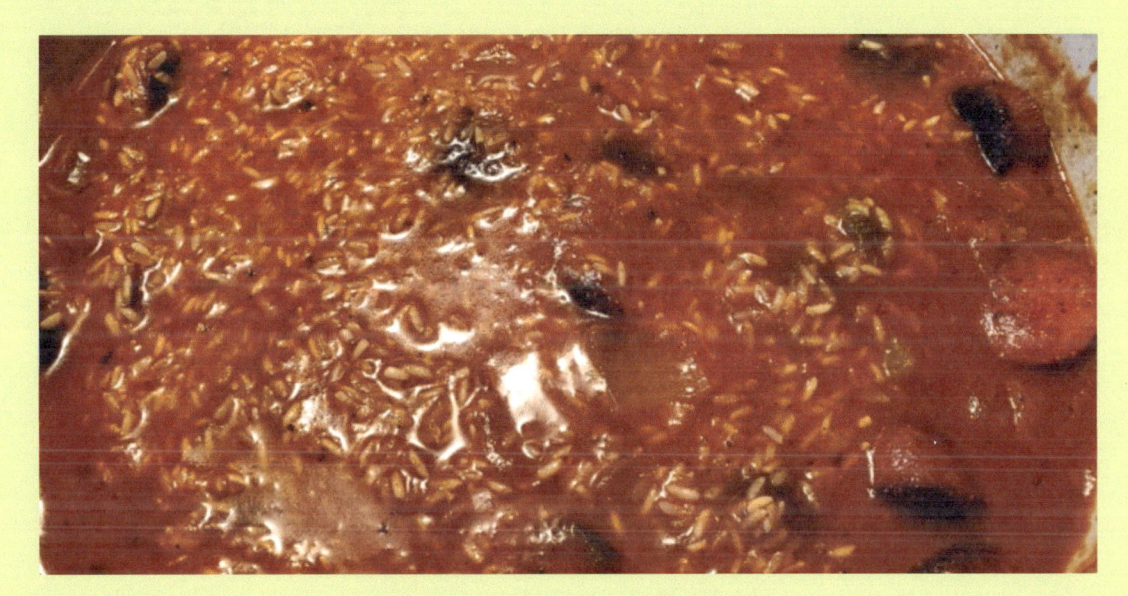

Every grain to itself or you did it wrong.

Red Rice for 'Comeyas'

Serves 8

2 tablespoon olive oil
1 pack smoked sausage, cut into thick slices
1 large onion, diced
2 stalks celery, diced
1 green pepper, diced
2 (14.5 oz.) cans crushed tomatoes
1 can tomato paste
2 tablespoons brown sugar
3 cloves garlic, chopped fine
1 teaspoon poultry seasoning
1 cup water or broth
¼ teaspoon black pepper
Uncle Ben's parboiled rice, to taste
1 teaspoon chopped fresh parsley

In large heavy pot, saute sausage until browned on all sides. Add oil if needed. Cook onions, peppers and celery in oil. Stir in garlic and poultry seasoning. Mix in tomato paste, brown sugar and tomato sauce until combined. Add water and bring to a boil. Add sausage back to the pot and cook until the sauce bubbles. Remove pot from heat and carefully pour in parboiled rice until there is just a thin layer of sauce remaining. Do. Not. Stir.

Cover pot with foil and bake at 350°F for 30-45 minutes.

NOTE: *Shout out to Akua (Geechee Experience) and the others wanting a vegan version! Skip the sausage and use oil to saute vegetable. Then, continue with instructions.*

10 Veggie Spaghetti
Serves 8 - 10

3 tablespoons oil
1 lb. ground beef, ground turkey, or Italian sausage
1 large onion, chopped
1 large carrot, diced
1 yellow pepper, chopped
½ bunch of kale, chopped (you can substitute whole spinach)
1 stalk of celery, chopped
½ pound of fresh mushrooms, chopped
1 squash, diced
1 zucchini, chopped
1 can olives, chopped
1 small eggplant, cubed
1 teaspoon each celery seed and garlic powder
½ teaspoon oregano or Italian seasoning
3 jars spaghetti sauce

Add oil to a large pot over medium heat, and sauté ground beef until browned. Add kale, carrots, onion, and celery, and cook for 10 minutes. Add peppers, mushrooms, squash, zucchini and eggplant, and cook for another 10 minutes. Stir in spices and add spaghetti sauce.

Serve hot over a bed of zucchini noodles, spaghetti squash, or traditional pasta.

NOTE: *This pasta sauce would also be great in lasagna or baked ziti. It also a good way to get kids to eat the whole rainbow over pasta.*

Lowcountry Steamed Crabs
Serves 3 - 5

1 quart of water or broth
1 can of beer
3 hot peppers, halved and seeded
5 cloves of garlic, peeled

2 onions, chopped
1 ½ cups of seafood seasoning
5 pounds of live crabs or crab legs

Add broth and beer to stock pot and bring to a boil. Season the pot with peppers, onions and garlic. Add seafood seasoning and hot peppers, bring to a boil. Add crab in batches, cover and steam for 5-10 minutes. Use tongs to remove crabs and repeat until all batches are steamed.

Eat straight on newspaper directly on the table and enjoy. Compost those crab shells and the spent newspaper, y'all!

NOTE: *The serving size depends on where you're from. It makes 5 servings if you're from out-of-town or 3 if you're from Charleston. Judge accordingly.*

Proud trainee after Chris Cato from Geechee Experience showed me how to '*buss em down.*'

Cookin' Youths

I'll never forget the day that I headed to Anik and Adrian's preschool for show and tell. Dressed up in my chef's coat and hat, I demonstrated how to construct fruit filled crepes drizzled with chocolate syrup. Each child constructed their own crepe. In a center for 4 and 5 year olds, Adrian was the youngest at 18 months (thanks to the late Ms. Robinson for taking him on). He was the last to prepare the dessert crepe and didn't miss a step. He's been cooking ever since.

We'll see if Glam-Baby can cook as well as Dadda.

Adrian's Chunky Chili

Serves 10

1 pound Italian sausage
1 pound ground beef
2 zucchinis, chopped
2 yellow squash, chopped
½ sweet onion, chopped
1 15 oz can of corn
15 oz salsa (mild)
48 oz pasta sauce
2 tsp sugar
2 bell peppers, chopped
¼ pound okra, chopped
2 - 15 oz cans black beans
2 Tbsp veg oil
Seasoned salt
Onion powder
Garlic powder
Black pepper
Italian seasoning
Cumin

Add vegetables oil to a large pot. Saute zucchini, yellow squash, onion, bell peppers and okra. Add italian sausage and ground beef. Add seasonings, stir. Add pasta sauce and salsa. Add corn and black beans.

NOTE: *This is my son, Adrian's, recipe, and it is GOOD! You're definitely gonna want seconds. Maybe even thirds. Fortunately, it makes a lot.*

Salmon Stew
Serves 6

2 tablespoons oil or bacon fat
1 large onion
2 cans salmon, skin and bone removed, liquid reserved

2 teaspoons of garlic powder
¼ to ½ cup ketchup
½ cup water

Heat the oil in a large saucepan over medium high heat. Add onions and cook until softened. Add salmon, garlic powder, and liquid from can, and cook until heated through. Add ketchup and water, and cook for another 5-10 minutes. Serve hot over rice.

My mom used bacon fat, but olive oil is a good substitute. FUN FACT: Bones in canned salmon can provide calcium. I liked to add them to the mix, the kids didn't love that idea.

French Pork n' Beans

We know about pork and beans. *Cassoulet* is the French version, *feijoada* is the Brazilian version (next cookbook). This is another favorite from my bougie food phase twenty five years ago. The traditional recipe takes at least 3 days to prepare, but this recipe makes it possible to pull together for Sunday dinner. Perfect stew for cold weather!

Great over rice or topped with buttered bread crumbs.

Quickie Cassoulet

Serves 8 - 10

4 bone-in, skin-on chicken thighs, cut in ½ through the bone
Salt and pepper to taste
½ pound slab bacon, cut into chunks
Smoked kielbasa
1 large onion, chopped
3 celery stalks, chopped
4 garlic cloves, diced
½ cup white wine
2 cans fire roasted tomatoes
3 cans Great Northern white beans
1 bay leaf
2 tablespoons dried thyme (or 3 fresh thyme sprigs)
½ cup chicken broth

Dry chicken thighs, season, and set aside. In a large pot, cook bacon chunks on medium low heat. Remove bacon when crisp, and raise heat. Saute the chicken on medium high heat, and brown on both sides. Remove chicken and brown sausage, celery, carrots, onions, and garlic in pan until vegetables soften. Add white wine, fire roasted tomatoes, beans, bay leaf, and thyme, and cook for 20 minutes, stirring occasionally to scrape the bottom of the pan. Stir in chicken and bacon into the bean mixture, and cover with chicken broth. Cover and bake at 350 degrees for 30 minutes.

Substitute fresh chopped tomatoes for the canned option and the polish sausage of your choice.

A Love for Cooking

When I was my eldest child's age, I dreamed of going to culinary school. Most of my spare money was spent buying cookbooks of the chefs I admired on PBS Saturday afternoon shows. The way other folks read romance novels, I was reading cookbooks from cover to cover.

My friend, Brian, bought me a Cuban cookbook, and a variation of this recipe was the first thing I prepared for my own baby shower. I didn't realize until a month ago that this is the Cuban version of red rice - YUM!

Arroz con Chorizo served with salad of cucumbers, red onions, black beans, arugula and chimichurri.

(G) (H) (DD)
Arroz con Chorizo
Serves 8 - 10

2 tablespoons olive oil
3 packages chorizo
1 green bell pepper, chopped
1 yellow bell pepper, chopped
3 stalks celery, chopped
1 large onion, chopped
5 cloves of garlic, minced garlic

4 cups white rice
2 cans chicken broth
½ cup white wine
⅛ teaspoon saffron
2 (14.5 oz.) cans stewed tomatoes, crushed
1 Tbsp chopped fresh parsley

Heat the oil in a large skillet over medium heat. Add chorizo, and cook for 8-10 minutes. Stir in pepper, onions, celery, and garlic, and cook for 7 more minutes. Add rice; cook and stir until rice is opaque, 1 to 2 minutes. Stir in broth, white wine, saffron, and tomatoes, and return to a boil. Stir in parsley, cover, and bake at 350 degrees for 20 minutes.

NOTE: *This is the Cuban cross between Red Rice and Chicken Bog.*

A bottle of dry white wine makes several batches of arroz con chorizo. I add a splash to stir fries too.

Veggies

Roasted vegetables are great on top of a cold salad.

Gullah Chef Benjamin BJ Dennis taught me that our ancestors cooked okra leaves and sweet potato leaves as their summer greens. Our team then showed farm campers how to prepare them over an open fire. You can too!

You can substitute squash or zucchini noodles for pasta in your soups.

Okra leaves chopped and ready to cook like collards.

(G) (VT) (VN) (30)
Roasted Broccoli
Serves 4

1 bunch broccoli (about 1½ pounds), cut into florets, stems peeled and sliced or diced

2 tablespoons extra-virgin olive oil
3 cloves garlic, sliced
Kosher salt and freshly ground pepper

Preheat oven to 450°F. Toss the broccoli florets with olive oil, garlic, salt, and pepper on a baking sheet. Spread them out and then roast, without stirring, until the edges are crispy and the stems are crisp tender, about 20 minutes. Serve warm.

NOTE: *Broccoli gets sweeter when roasted.*

Drizzle with lemon juice for a special treat

Lowcountry Soup Bunch

(G) (VT)

Serves 8

2 Tbsp olive oil
1 medium onion, chopped
5 carrots, peeled and chopped
3 stalks celery, chopped
4 cloves of garlic, minced
4 cans low sodium chicken or veggie broth
2 cans diced tomatoes (don't drain)
3 medium potatoes, peeled and chopped into ½ inch thick pieces
⅓ cup fresh parsley, chopped (optional)
2 bay leaves
½ tsp. Dried thyme
Salt and pepper, to taste
1½ cups green beans (fresh or frozen)
1½ cups corn (fresh or frozen)
1 cup peas (fresh or frozen)

Heat olive oil in a large pot over med-high heat. Add garlic, onion, carrots, and celery and saute until onions are translucent. Add broth, diced tomatoes, potatoes, parsley, bay leaves, and seasoning. Bring to boil. Reduce heat and bring to simmer until potatoes are almost soft. Add green beans, corn, peas and cooked until they are warmed through. Serve and enjoy!

NOTE: *The secret to this is basically throwing all your leftover veggies in a pot. Do that and you got it.*

Diced cabbage, rutabaga, and turnips are also delicious in your soup bunch.

Okra Soup with Oxtails
Serves 8

3 tablespoons olive oil
3 lbs oxtails or 1 lb stew beef, rinsed and dried
1 teaspoon garlic powder
1 teaspoon poultry seasoning
Salt and pepper to taste
1 tablespoon butter
4 green onions (thinly sliced, white and green parts separated)
1 small clove garlic (minced)
1 rib celery (sliced)
2 cups chicken stock (preferably unsalted)
4 large tomatoes, broiled and rough chopped or 1 (14.5-ounce) can diced tomatoes (undrained)
1 lb cut okra (about 3 cups sliced, fresh or frozen)
1 cup corn kernels (fresh, canned, or frozen)
1 teaspoon Creole seasoning
Salt (to taste)
Pepper (to taste)

Season the oxtails, and set aside. Add oil to a large saucepan over medium-low heat, and brown the oxtails on all sides. Remove oxtails, and stir in chopped onions, celery. Add the white parts of the green onion, garlic, and celery; sauté until celery is tender. Add chicken stock, tomatoes, sliced okra, corn kernels, and Creole seasoning.

Bring the soup to a boil. Reduce the heat to low, and cover the pan. Simmer for 20 to 30 minutes, or until the okra is tender.

Taste and season with salt and pepper, as desired. Sprinkle servings with the reserved green onion tops.

Teriyaki Eggplant w/ Spaghetti Squash

(G) (VT) (VN)

Serves 4

For the Eggplant:
- 6 small or two large eggplant, diced
- 1 onion, diced
- 3 Tbsp olive oil
- garlic powder, to taste
- bottle of teriyaki sauce

For the Spaghetti Squash:
- 1 spaghetti squash
- 1 Tbsp olive oil
- garlic powder, to taste

Place whole spaghetti squash in baking dish. Bake squash at 350 degrees for 45 minutes to an hour. Remove from oven and cool for at least 15 minutes. Use a sharp knife to cut squash in half. Remove seeds with a spoon. Shred squash into a large bowl. Toss with olive oil and garlic powder while still warm. Set aside.

Saute eggplant and onions in oil to desired texture. Stir in garlic powder. Add teriyaki sauce and heat through. Serve and enjoy!

(G) (VT) (VN)
Chef BJ's Eggplant Creole
Serves 6

5 Tbsp olive oil, divided
2 medium eggplants, chopped
1 large onion, chopped
3 stalks celery, chopped
1 red bell peppers, seeded and chopped
3 cloves garlic, minced
1 teaspoon fresh ginger, minced
1 tsp dried thyme
1 tsp salt
¾ tsp black pepper
1 tablespoon tomato paste
2 15 ounce cans crushed tomatoes or 4 medium beefsteak tomatoes, diced
1½ cups low sodium vegetable broth
2 tablespoons fresh parsley, chopped

Coat eggplant, bell pepper, and onion in 2 tablespoons of olive oil. Season with salt and pepper. Spread a single layer of veggies, and broil for 3-6 minutes, tossing halfway through cooking. Continue roasting veggies in batches. BE SURE TO WEAR OVEN MITTS!!! Heat remaining olive oil in a large pot over medium heat. Add roasted vegetables and celery, and cook celery and roasted onions until they are softened. Stir in tomato paste, then add garlic, ginger, and cook for 3-5 minutes. Add tomatoes, broth, and cover. Reduce the heat to low, and simmer for about 20 minutes. Stir in parsley, and serve hot.

We don't taste bitterness in farm fresh eggplant.

Veggied Lima Beans

Serves 8

3 tablespoons oil
1 large carrot, diced
1 large onion, chopped
1 bell pepper, seeded and chopped
1 yellow squash, chopped
1 zucchini, chopped
4 cloves of garlic, chopped
½ teaspoon of celery seeds

1 teaspoon of dried thyme
1 lb dried baby lima beans, soaked in hot water for an hour and rinsed
2 cups vegetable or chicken broth
1 bay leaf
2 teaspoons salt
1 tablespoon hot sauce
Black pepper to taste

Add oil to a large saucepan and saute vegetables until browned. Stir in garlic and seasonings. Add beans, broth and bay leaf and bring to a boil.

Reduce heat and simmer on low heat for 45 minutes to an hour, stirring frequently. Add water as needed to keep beans moist until they soften. When liquid reaches gravy like thickness the beans are done.

Re-season with garlic powder, and add salt, pepper and hot sauce. Serve warm.

NOTE: *You can substitute black-eyed peas or field peas for the lima beans, and it's still delicious.*

Desserts

Add a teaspoon of cinnamon to chocolate cake batter.

Replace water in a cake recipe with yogurt or sour cream for a texture similar to pound cake.

Mix ½ cup to a cup of rolled oats in your batters to increase the fiber without changing the taste.

Make your own brown sugar with molasses and sugar.

DIY brown sugar by incorporating a teaspoon of molasses into a 1/4 cup of sugar.

Dump Cake
Serves 12

1, 21-oz. can cherry pie filling
1, 15-oz. can crushed pineapple
1 box white cake mix (18 oz.)
1 cup rolled oats½ pecans or walnuts, chopped
1½ sticks (12 tablespoons) butter

Dump the cherry pie filling followed by crushed pineapple into a 9-by-13-inch baking dish. Stir together. Mix rolled oats with cake mix, and sprinkle over fruit. Slice the butter into tablespoons, and distribute evenly over the surface of the cake mix. Top with chopped nuts. Bake at 350 degrees until the tops are brown and bubbly, 45 minutes to 1 hour.

NOTE: *Here are 3 other filling options I've tried and approved.*
- *Blueberry - 4 cups fresh blueberries and ½ cup sugar*
- *Apple - 6 apples, peeled, cored, and thinly sliced, ½ packed brown sugar*
- *Cranberry - 1 can whole berry cranberry sauce and 1 cup of berries*

Rolled oats add fiber and a nice chewy texture.

Sweet Potato Bread Pudding
Serves 8

6 cups torn whole wheat bread
4 Tbsp butter, melted
1 tsp ground cinnamon
1 cup light brown sugar (packed)

2 ½ cups almond milk
3 large eggs
1 ½ tsp vanilla
2 cups mashed sweet potatoes

Butter a 2 quart baking dish. In a large bowl, whisk together milk, eggs, and brown sugar. Then, add the vanilla, and blend with the sweet potatoes. Toss whole wheat bread into the sweet potato mixture, and let stand for 30 minutes. Transfer to the buttered baking dish, and bake at 350°F for 45-55 minutes or until the mixture is set. Let cool and serve.

I substituted canned pumpkin and Dave's Killer bread for a fancy grand opening.

Vegan Mousse
Serves 8

1 can coconut cream
1, 10 ounce bag vegan chocolate chips (Enjoy Life, Whole Foods 365 Brand, etc.)

1 teaspoon cinnamon
½ cup toasted hazelnuts

Shake cans until coconut cream blended then open. Warm over medium heat just until warmed through. Stir in chocolate chips and cinnamon. Add foil muffin cups to a muffin pan and fill ¾ full with chocolate mixture. Refrigerate for an hour or until it sets. Top with hazelnuts and serve.

NOTE: *This recipe was inspired by a cooking demo by Chefarmer Matthew Raiford and Jovan Sage at a Charleston Wine & Food event.*

Briana became an FFF ambassador after she made and tasted that mousse! Not a coincidence, y'all...

Coconut Pecan Cookies

Makes about 20 cookies

1 cup brown sugar
1 cup white sugar
1 cup butter, softened
2 eggs
1 tsp vanilla
1½ cup flour

1 tsp salt
1 tsp baking soda
1½ cup uncooked oats
2 cups coconut flakes
1 cup chopped nuts

Preheat the oven to 350°F. In a small bowl, cream butter and sugar. Beat in eggs and vanilla. Set aside. In a large bowl, sift dry ingredients, mix together. When evenly mixed, add in the wet ingredients. Drop spoonfuls of dough on a greased cookie sheet and bake for 8 minutes. Repeat until all the batter is gone or freeze the leftovers for another day.

NOTE: *My eldest, Anik, was the cookie chef in our house. If we wanted cookies, we'd ask them. This is by far one of our family favorites. They're also great as bar cookies.*

Pecans from the one established tree on the Farm

(G) (H) (VT)

Peach Charlotte
Serves 6

Another magnificent and simple dessert, I discovered while doing interpretive work with my buddy, Jerome, at Magnolia Plantation.

1 loaf white bread, crusts removed
3 sticks of butter, softened
¾ cup sugar
1 teaspoon cinnamon
2 cans of peaches, drained
2 dashes each nutmeg and mace

Butter a casserole dish and set aside. Mix sugar with spices and set aside. Coat all bread with butter on both sides. Line the bottom and sides of the casserole dish with bread. Sprinkle the spiced sugar on the bread, add a layer of peaches, and then sprinkle with the sugar mixture. Repeat with a layer of bread, add peaches, sugar mixture, and finish with bread. Cover and bake at 350°F for 40 minutes or until golden brown.

Butter, bread, peaches, and spices make magic!

Teas & Tonics

Pour your favorite cold tea into a soda stream for a carbonated treat.

Save the bottoms from your lemongrass to root and grow your own!

Add natural flavor and sweetness to your tea with sliced citrus and berries.

Easy chai tea = simmer a black tea bag in almond milk and add a dash or three of pumpkin pie spice.

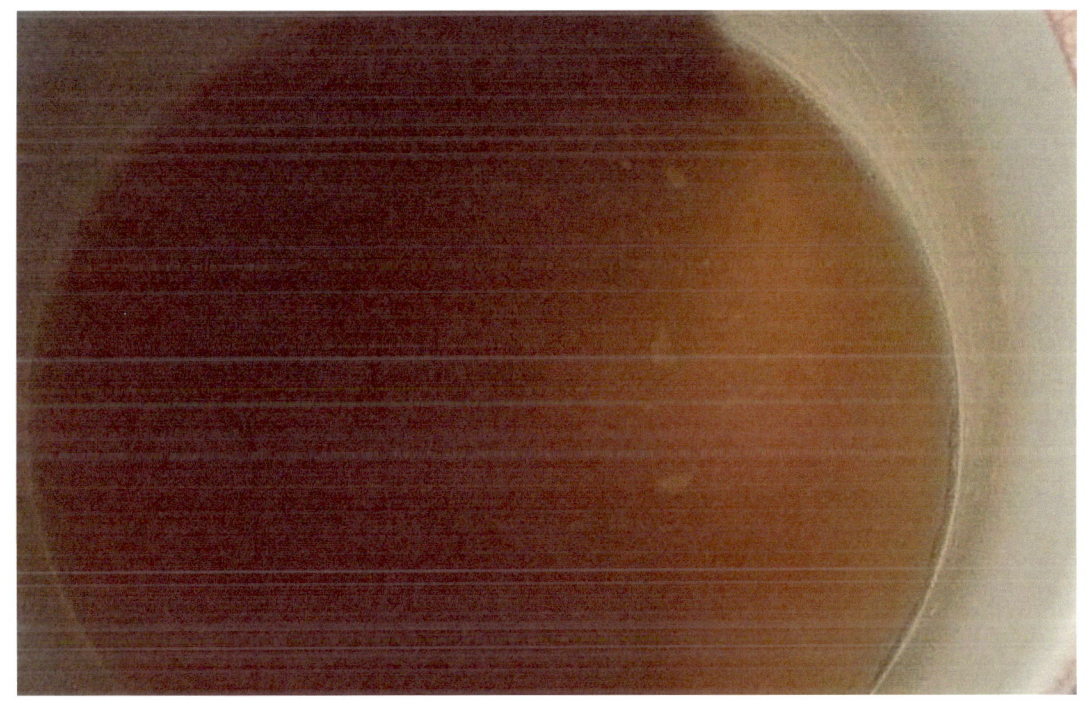

An apple honey tasting batch Anik prepared

Lavender Lemonade
Serves 10

Simple Syrup:
- ¼ cup dried (or a bunch of fresh) lavender
- 3 cups boiling water
- 1 cup white sugar

Lemonade:
- 8 lemons
- 4 cups cold water, or as needed

Heat water to boiling in a medium pot. Add lavender, and steep on low heat for at least 30 minutes. Strain out the lavender and discard. Mix the sugar into the hot lavender water, then pour into a pitcher with the ice. Juice the lemons into the pitcher. Add cold water and stir.

NOTE: *I have to thank Karen Latsbaugh with Cities + Shovels for this now farm camp favorite (and for introducing me to gardening many moons ago).*

Mountain Mint Tea
Serves 2

2 cups water
4 teaspoons sugar
15 mountain mint leaves

Bring water to a boil. Remove from the heat and add mint leaves. Let steep for 3 to 5 minutes. Add sugar and enjoy!

Fresh Herb Sun Tea
Makes 2 Quarts

2 quarts water
¼ cup wildflower honey
1 large lemon

2 bunches fresh herbs (mint, lemongrass, basil, etc.)

Pour the water into a large glass container with a lid. Stir in honey until dissolved. Zest the lemon, being careful not to include any white pith, and juice the lemon. Combine the zest and juice with honey water mixture. Gently bruise the herbs (to release oils and fragrance). Press the bunches in the water mixture. Cover the container and place in direct sunlight for two hours. Remove the mint leaves, shake, and serve over ice in tall glasses, garnished with a mint sprig.

Ginger Tea
Serves 4

5 cups water
3 one inch chunks fresh ginger, mashed with a wooden spoon

Honey to taste
1 fresh lemon or lime, juiced

Bring water to a boil in a small saucepan; add ginger and simmer on low for 20 minutes. Strain tea into a large glass; add honey and lemon wedge.

NOTE: *I make fresh ginger tea to settle my stomach.*

Peach Tree Dance

In April of 2018, an African dance and drum crew performed a West African harvest dance at the farm. By June, we harvested the first crop of the best peaches we ever tasted in our lives. That isn't a coincidence folks. The peaches off the tree had enough juice to fill a cup but our friend, Gullah Chef Benjamin 'BJ' Dennis told me we could make tea with the pits. That summer, we prepared peach tea chilly bears with fresh fruit as a frozen treat for farm campers.

These are peaches from the farm's tree on display for sale in the farm store.

Peach Pit Tea

Makes 2 Quarts

2¼ cups water
1 handful peach pits

6 cups cold water
1½ cups sugar or honey

Steep peach pits (about 1 per cup of tea) in 2-¼ cups boiling water. For extra flavor, steep them overnight. Discard the pits and heat the peach-flavored water to a gentle boil. Add sugar or honey until it dissolves. Serve hot. Or let cool, and pour over ice.

NOTE: *You can preserve the pits by drying them in your oven at 200 degrees for an hour. Store them in a tightly closed glass jar. This way you can enjoy peach pit tea even when peach season's over. You can collect fresh pits in the refrigerator until you're ready to preserve them.*

Peach tea with chunks of frozen farm peaches make a great chilly bear.

Hibiscus Tea Drink

A Chicora/Cherokee neighbor brought us our first hibiscus plant seedlings. The buds from the plant make a delicious Caribbean drink known as *sorrel* and *bissop*. It's called *jamaica* in Spanish. Anyway, the drink is normally consumed as a holiday beverage, usually spiked with rum. Fortunately for us, Bonita Clemmons from Columbia, SC supplies our farm store year round.

FFF Product of the Week:

BONITA'S HIBISCUS TEA

South Carolina's own Bonita's hibiscus tea for sale in our farm store!

(G) (VT) (VN)
Bissop, Sorrel, or Hibiscus Tea
Serves 5 - 6

10 cups of water or more (this amount can be adjusted for taste)
¼ cup ginger, mashed with a wooden spoon
8-10 cracked pimentos
1 lime chopped
1 cinnamon stick
2 cups of fresh or dried red sorrel buds
Simple syrup (1 cup sugar to two cups of hot water)

Bring water to a boil in a large pot. Add ginger, pimentos, lime, and cinnamon to pot. Bring to a boil. Add hibiscus buds, and simmer for another 45 minutes. Use a strainer to remove all solids. Add simple syrup to taste.

Thanks to our neighbor who provided the first hibiscus seedlings and a bountiful harvest in 2018

Allergy Brew for my Eldest

Anik developed severe food allergies in high school. After years of taking over the counter medications, we learned they could cause bodily harm from long term use. We discovered that kombucha relieved Anik's allergy symptoms. They found a recipe in college and learned to make it from scratch, which is a much cheaper proposition.

Anik reused kombucha bottles. They filled them with their own kombucha for easy portions and storage.

(G) (DD) (VT) (VN)

Anik's Kombucha
Makes 1/2 Gallon

4 black tea bags
½ cup sugar
6 - 7 cups water

1 bottle of ginger or plain kombucha, room temperature or scoby
½ - 1 cup fruit juice, optional

In a large pot, simmer water over medium low boil in a small pot. Add sugar and stir until it dissolves. Steep tea bags in the simple syrup for 10-15 minutes or until the water gets very dark. Cool the mixture to room temperature. Remove the tea bags, and pour the tea into a half gallon jar. Add a bottle of room temperature kombucha. If you can't find kombucha, a scoby can be purchased online. Cover the jar with a tight-weave towel or coffee filter, and secure with a rubber band. Allow the mixture to sit undisturbed at room temperature, out of direct sunlight, for at least 4-6 days. The longer the kombucha ferments, the less sweet and more vinegary it will taste.

Transfer finished kombucha to another jar. Reserve the scoby with enough liquid to cover to start your next batch of tea. If you prefer plain kombucha, you can refrigerate at this point. For a flavored batch, add ½ to 1 cup of fruit juice, and ferment for two more days. Then refrigerate.

NOTE: *For a lighter taste, you can make jun. It's similar to kombucha but uses green tea instead of black and honey instead of sugar.*

Acknowledgements

Cusabo Native Americans who first inhabited the region that includes North Charleston, SC

Along with my family --- Joan Hall (any grit in me is inherited from my mom), my late father, Willie Hall Sr. (all my silliness and spirit for justice came from him), brothers Illya and Will, and my sister-in-love Nikki Johnson --- and our Fresh Future Farm staff, there are so many people that helped me grow as a person and cultivate the idea of Fresh Future Farm.

FFF Co-Founder/Collaborator-in-Chief: Todd Chas

Advisors: Jennet Alterman, Sharon Funderburk, and Stacia Murphy

Fresh Future Farm board members (past and present): Karen Mae Black, Shaneak Brown, John Bukofser, Catherine Heston, Karen Latsbaugh, Mahwish 'Mev' McIntosh, Kennae Miller, Lee Moultrie, Barbara Nwokike, Angenita Owens, Angie Quirk-Garvan, Steve Saltzman, and Jara Sturdivant-Wilson.

Special Thanks: Emily Abedon, Andre Ameer, Alex Amit, Amerihealth Caritas Partnership, Anonymous Donors, Ed Astle, Bridget Besaw, Jerome Bias, BoomTown, Blue Pearl Farms, Councilman Michael Brown (District 10), Jessica Boyleston, Charlotte Caldwell, Chris Carnavale, Chad Carter, Charleston Magazine, Charleston Wine & Food, Chris Cato, Harry Chrissy, Clemson Extension Tricounty Master Gardeners, Coastal Community Foundation, Comcast CARES, Amy Dabbs, Katie Dahlheim, Sara Daise, Benjamin 'BJ' Dennis, Susan DuPlessis, Cara Ernst, Paul Garbarini, Greg Garvan, Gildea Foundation, Ian & Kimberly Gleason, Jonathan Green, Pastor Wendy Hudson-Jacoby, Carolyn Hunter-Heyward, Stephanie Hunt, Julie Hussey, Carolyn Lackey, Tim Latsbaugh, Mary Lynn Kohl, Kaye & Randy Koonce, Steven Kleiman, Andrea Limehouse, Jonell Logan, Dr. Kim Long, Lowcountry Alliance for Model Communities, Lowcountry Blessing Box, April Magill, Bernie Mazyck, Robin McCoy, Joseph McGill, Metanoia,

Marcus Middleton, Kennae Miller, Nicole Moore, Moss & Yantis, Omar Muhammad, Deljuan Murphy, Stacia Murphy, NeighborWorks America, Zsofia & Zsolt Pasztor, Angie Pitts, Brady Quirk-Garvan, Matthew Raiford, Ray Raiford, Fillipo Ravalico, David A. Root, Esquire, Roper Foundation, Jovan Sage, Jerry Scheer, Margarita Schmid, Les Schwartz, Select Health of SC, Darcy Shankland, South Arts, South Carolina Community Loan Fund, SC Arts Commission, South Carolina Association for Community Economic Development, Spaulding Paolozzi Foundation, Dean Stephens, Lindsay & Henry Street, Sandy Tecklenburg, Nick Tittle, Ben Towill, The Urban Electric Co., Svenja Xeller, Dan Xeller, Select Health of SC, David Walters, Sr., Sustainable Warehouse, Ali Titus, Gilbert Walker and Dontavius Williams

College of Charleston Interns: Victor Bennett, Laura Chicuazuque, Carly Dale, Garneisha Pinder, Susannah Rogers, Tiffany Singleton, and Johnsie Wilkinson

Farming/Gardening Innovators & Social Justice Pioneers: Will Allen, Erika Allen, George Washington Carver, Masanobu Fukuoka, Muhiyyidin D'baha (Moya Moye), Fannie Lou Hamer, Leah Penniman, Laurell Sims, Ruth Stout, and Malik Yakini